Early Childhood Numeracy for the Caribbean

123 You and Me

Activity Book 1

Weida Billings Whitbourne

Advisor: Cathryn O'Sullivan

Introduction

Developed in accordance with curriculum from across the region, *1, 2, 3 You and Me* is a complete package of interactive materials for Early Childhood Numeracy education in the Caribbean. It is child-centred, play-based, hands-on, developmentally appropriate and integrated. This package will help make teaching and learning enjoyable both at home and at school - everything you need to teach and excite young learners!

The *1, 2, 3 You and Me* Activity Books have been designed to promote Early Childhood Numeracy in a fun and interactive way. They can be used in conjunction with the full *1, 2, 3 You and Me* package, or can be used separately.

Activity Book 1 introduces foundational Numeracy skills through a range of activities that have a familiar sequence but become more challenging as the book progresses up to numeral 10.

Activity Book 2 builds on the skills that were introduced in Activity Book 1 by addressing the same material in a more complex manner and by introducing new concepts such as simple addition, as well as extending activities up to numeral 15.

Allow children beginning to develop their Numeracy skills to have fun and express themselves in these Activity Books - both books are designed to fit with and complement wider classroom or home activities, and encourage children to develop at their own pace. See how the teaching and learning of Numeracy can be enhanced with the full *1, 2, 3 You and Me* pack components, listed on the next page.

Material is ordered by numerals and interspersed with other Numeracy topics – each page has its own heading to outline what is being covered

'Looking back' pages are included throughout to help with the revision of topics as children progress

Each Activity Book includes lots of opportunity for children to draw and colour in

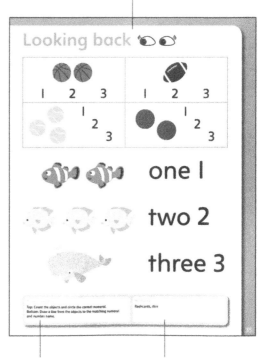

Pages introducing numbers include tracing activities with additional blank boxes for children to write the numerals if they are ready. If children are not ready to write, theses boxes can be used for additional tracing practice or be left blank

Activity instructions at the bottom of each page are there to guide the activity and should fit in with your approach to class/home discussions and other activities

Suggestions for incorporating hands-on activities using *1, 2, 3, You and Me* pack components, including the Teacher's Guide, are provided in the 'Resources' box at the bottom of each page. Be creative about using the materials

The *1, 2, 3, You and Me* package

Along with the **Activity Books**, here is what is included:

Teacher's Guide supports the full package with strategies to adapt regular classroom activities, including conversations to encourage critical thinking, activities for home and school, environmental observations, as well as a special focus on parental involvement and approaches to teaching children with special needs.

Parrot puppet is provided as a fun and engaging means of interacting with children. Its character is used throughout the books; use the puppet to introduce topics and with specific activities found in the Teacher's Guide as well as in the Activity Books. Let children give him or her a name!

Reward stickers - encourage and support children's efforts with reward stickers.

Flashcards are provided for numbers 0-12. Four sets of flashcards are provided to facilitate independent use as well as the playing of games like "go fish", "memory" or matching. The numeral and number name are on the front with quantities represented on the back. The arrangement of the pictures helps to introduce the concept of odd vs even, and the repeated background colours offer opportunity for colour recognition and matching games. Use the flashcards to complement the Activity Books, and learn more about additional ways to use them in this Guide.

Numeral Frieze is brightly coloured and includes numerals 0-10. It's the perfect addition to the classroom wall. As with the flashcards, children can continue to learn about numerals, number words, colours and the concept of odd versus even.

Audio CD-ROM - have fun singing with your class with Numeracy-related songs and stories. Song lyrics

and a numeral formation guide for the number formation song are provided in the Teacher's Guide.

Board Games - Four board games are included: (1)The Squeeze, (2) Colourful Shapes, (3) Snakes and Ladders and (4) The Big Race. These promote the development of Numeracy skills, and as group games they encourage teamwork and sportsmanship. They can be used in conjunction with the dice, stickers and Unifix cubes. Detailed instructions for the board games, as well as suggestions for additional games, can be found in the Teacher's Guide.

Counter stickers are included for use with the games and activities including counting, sorting and matching. Use them on the top of everyday items like plastic bottle caps.

Unifix cubes - one set of 100 colourful, interlocking counting cubes represent 'units' and can be used in conjunction with the board games, Activity Book activities and additional games suggested in this Guide.

Dice with numerals and quantities are provided primarily for use with the board games. Children will also be able to start to associate the quantity with the numeral and can count the dots for verification. Suggestions for other activities involving dice can be found in this Teacher's Guide and the Activity Books.

Tangrams are an ancient Chinese puzzle comprised of seven shapes. Two sets of colourful tangrams are provided in this pack. They are versatile and can be put together in hundreds of ways to make a variety of shapes including people and animals. A detailed description of tangrams and how to use them is provided in the Teacher's Guide.

And everything can be stored in the reusab *1, 2, 3, You and Me* drawstring bag!

> Children are just beginning to develop Numeracy skills and it is important to encourage and foster their love of learning. Remember to praise them and their efforts and when checking for accuracy focus on the aim of the activity rather than their drawing skills.

Contents

One-to-one correspondence

Activity instructions

Draw a line from 1 animal to 1 object.

Resources

flashcards, frieze
Teacher's Guide page 5

Red

Activity instructions
Talk about things that are red. Colour the pictures red.

Resources
counters with red objects, red cubes, songs: Red, Colours
Teacher's Guide page 11

2

Circles

Activity instructions

Talk about circles and things that are shaped like a circle. Colour the circles red.

Resources

Colourful Shapes game, shape counters, songs: Red; Colours; Shape song
Teacher's Guide page 10

One 1

Activity instructions
Colour the picture of the numeral 1 and the apple.

Resources
flashcards, frieze, songs: Little fingers; Number formations
Teacher's Guide page 9

One 1

I

I

Activity instructions

Top: Talk about the kitten.
Middle: Trace the numeral 1.
Bottom: Colour 1 kitten and the basket.

Resources

flashcards, frieze, songs: Little fingers; Number
formations; One, two, buckle my shoe

One 1

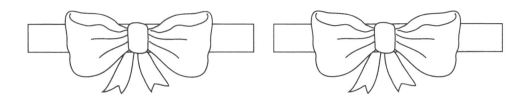

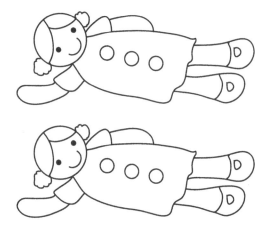 

Trace						

Activity instructions

Look at the objects. What can you see? Colour 1 object in each set.

Resources

flashcards, frieze, songs: Little fingers; Number formations;
One, two, buckle my shoe

One I

Trace						
	I	I	I	I		

Activity instructions

Look at the fruit. Draw lines from the sets of 1 to the numeral 1 and colour.

Resources

flashcards, frieze, songs: Little fingers; Number formations; One, two buckle my shoe

One 1

Trace	⋮	⋮	⋮			

Activity instructions

Help Lauren celebrate her birthday. Cut and paste or draw 1 candle on the birthday cake. Colour the birthday cake.

Resources

flashcards, frieze, songs: Little fingers; Number formations; One, two, buckle my shoe

Taller

Shorter

Green

Activity instructions

Talk about things that are green. Colour the pictures green.

Resources

parrot puppet, green cubes, counters with green objects,
song: Ten green bottles
Teacher's Guide page 11

11

Two 2

Activity instructions

Make a swan out of the numeral 2. Draw a beak, eye, wing and water underneath the swan. Colour the swan.

Resources

flashcards, frieze, songs: Number formations; One, two, buckle my shoe
Teacher's Guide page 9

Two 2

1   2

2 2 2 2 2

Two 2

Trace	2	2	2	2		

Two 2

2

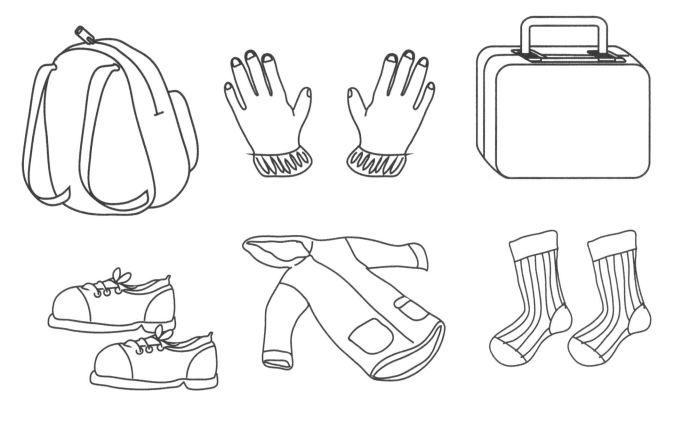

Trace	2	2	2	2		

Two 2

Trace	2	2	2	2		

Two 2

2

Activity instructions
Top: Draw 1 more apple.
Middle: Draw 1 more flower. Trace the numeral 2.
Bottom: Draw 2 leaves.

Resources
flashcards, frieze, songs: Number formations; One, two, buckle my shoe; Little fingers

Looking back

1				
2	2	2		

Activity instructions

Talk about the features that make up your face. Draw your face in the mirror. Remember your 2 eyes, 2 ears, 1 nose and 1 mouth. Colour your picture.

Resources

flashcards, frieze, songs: Number formations; Little fingers

Sets

Activity instructions

The parrot is hungry. Draw 2 things on the plate for him to eat.

Resources

parrot, counters, flashcards, frieze
Teacher's Guide page 6

Same

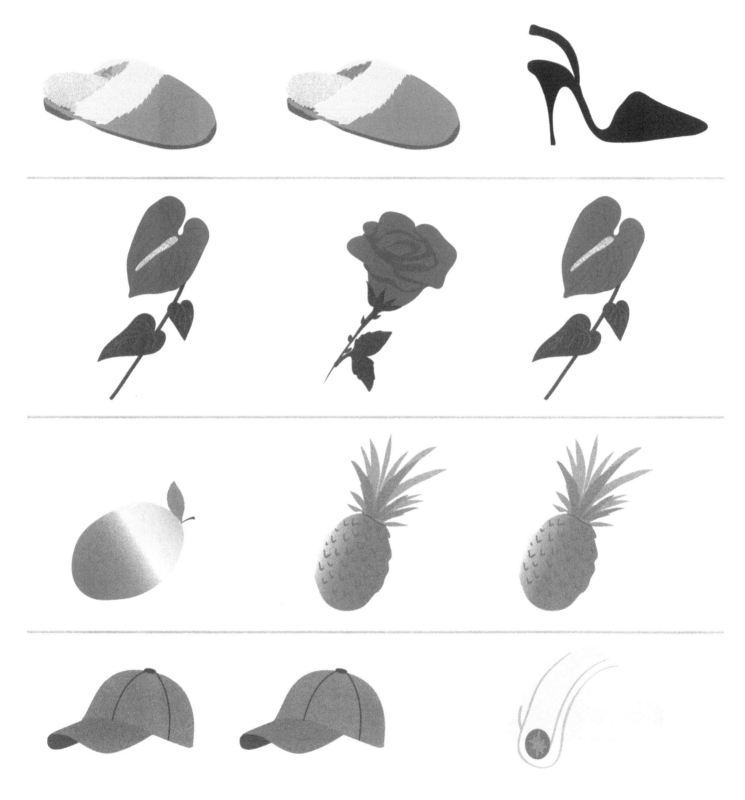

Look at the pictures. What can you see? Draw a circle around the things that look the same in each line.

counters, cubes
Teacher's Guide page 16

Different

Activity instructions

Look at the pictures. What can you see? Draw a circle around the picture that is different in each line.

Resources

counters, cubes
Teacher's Guide page 16

Yellow

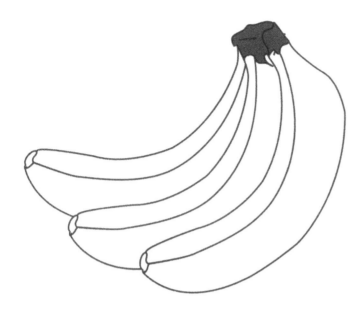

Activity instructions	Resources
Talk about things that are yellow. Colour the pictures yellow.	flashcards, counters with yellow objects, yellow cubes, song: Colours Teacher's Guide page 11

22

Triangles

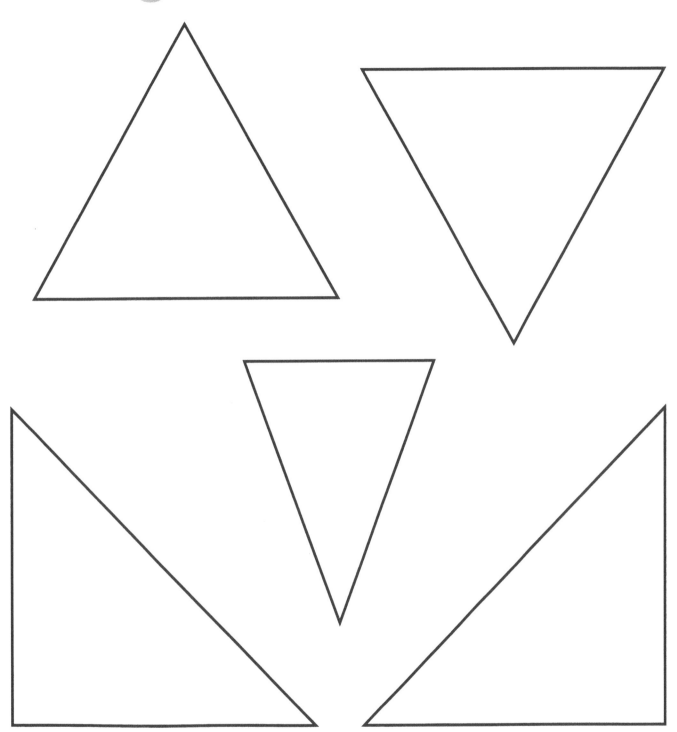

Activity instructions

Talk about triangles and things that are shaped like a triangle. Colour the triangles yellow.

Resources

Colourful Shapes game, shape counters, tangrams, song: My hat it has three corners
Teacher's Guide page 10

23

Three 3

Activity instructions

Tear coloured paper and paste it on the numeral 3.

Resources

flashcards, frieze, songs: One, two, buckle my shoe; Little fingers; Three little kittens

Teacher's Guide page 9

Three 3

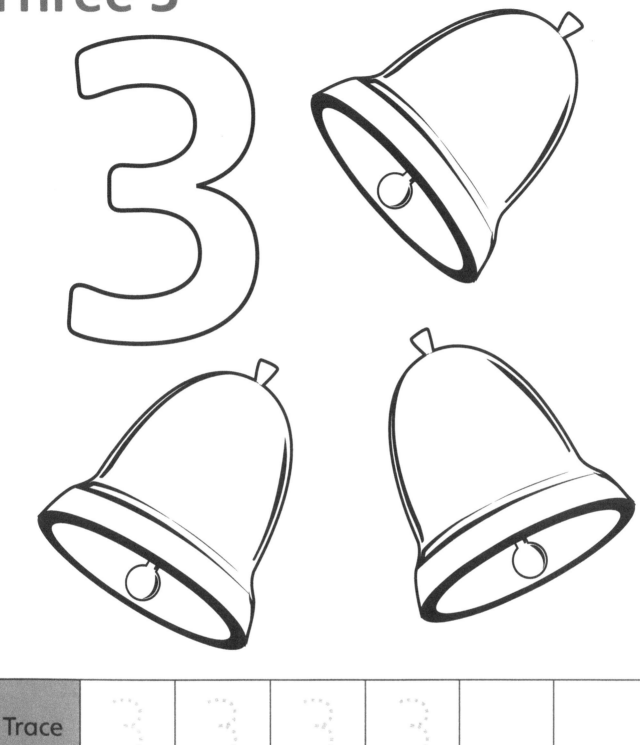

Trace	3	3	3	3		

Activity instructions

Top: Count and colour the bells.

Bottom: Trace the numeral 3.

Resources

flashcards, frieze, song: Number formations

Three 3

1   2   3

3   3   3   3   3

RUFF

Activity instructions

Top: Count the dogs.
Middle: Trace the numeral 3.
Bottom: Find and colour 3 balls.

Resources

flashcards, frieze, counters, songs: Number formations;
Three little kittens

Three 3

Trace						

Activity instructions
Top: Stick coloured sand or tear and glue paper to make 3 scoops of ice cream. Colour the cone.
Bottom: Trace the numeral 3.

Resources
flashcards, frieze, counters, dice, cubes

Three 3

Activity instructions

Look at the pictures. Count and colour all sets of 3.

Resources

story: The three little pigs

Three 3

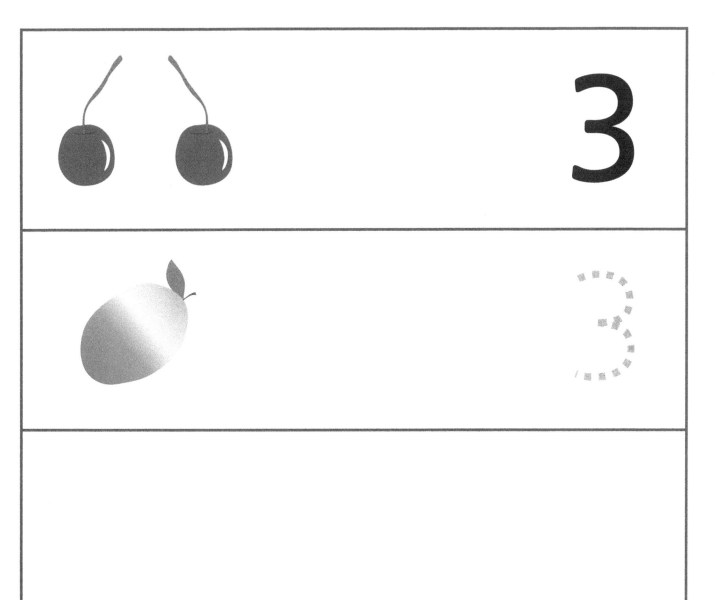

Activity instructions

Top: Draw 1 more cherry.
Middle: Draw 2 more mangoes.
Bottom: Draw 3 balls. Trace the numeral 3 and the number name.

Resources

flashcards, frieze, counters, dice, cubes
song: Number formations

Longer

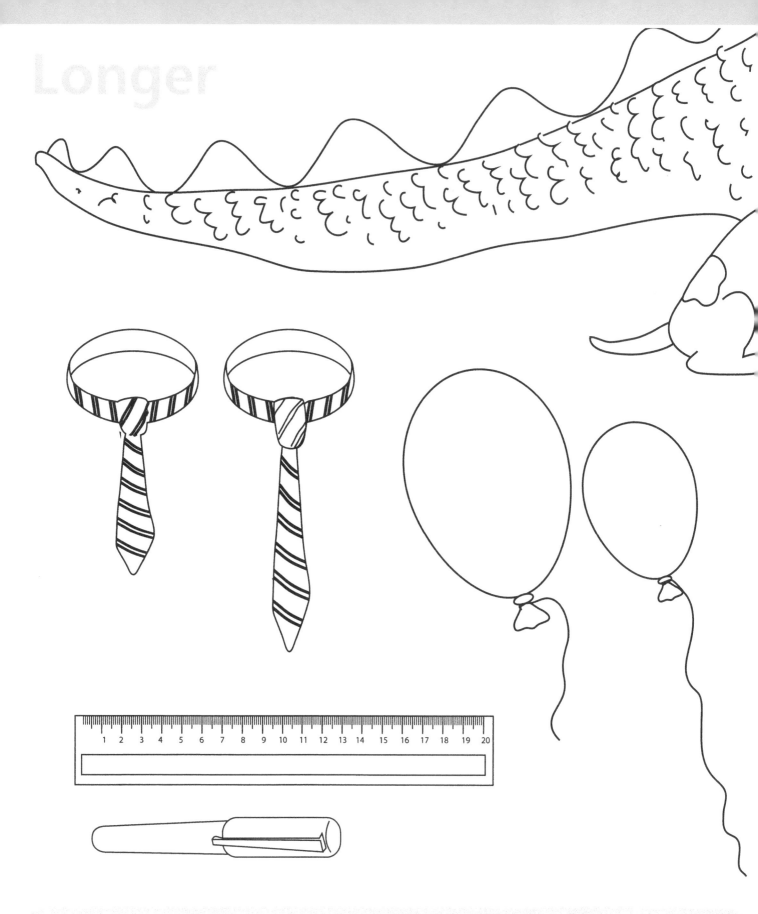

Look at each set of pictures. Colour the longer thing in each set.

cubes
Teacher's Guide page 13

Shorter

Activity instructions

Look at the sets of pictures. Circle the shorter thing in each set.

Resources

cubes
Teacher's Guide page 13

Colour by numbers

orange 1

yellow 2

blue 3

Activity instructions
Look at the duck. Colour: 1 orange, 2 yellow, 3 blue.

Resources
duck counter stickers, song: Five little ducks
Teacher's Guide page 11

Looking back

1

2

3

Activity instructions
Read the number in each box. Draw pictures to match the numeral.

Resources
flashcards, frieze, dice, song: Little fingers

33

Looking back

3	
2	
1	

Activity instructions

Top: Look at the numeral and colour the same number of cats.
Middle: Look at the numeral and colour the same number of dogs.
Bottom: Look at the numeral and colour the same number of chicks.
Draw three spots on the dog and colour the dog.

Resources

flashcards, frieze, dice, song: Three little kittens

Looking back 👁️👁️

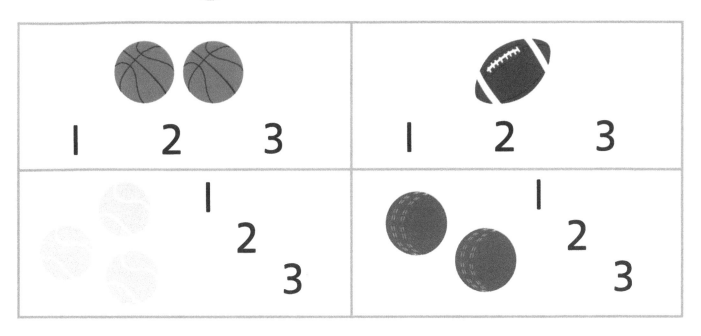

1 2 3	1 2 3
1 2 3	1 2 3

one 1

two 2

three 3

Activity instructions

Top: Count the objects and circle the correct numeral.
Bottom: Draw a line from the objects to the matching numeral and number name.

Resources

flashcards, dice

35

Blue

Activity instructions

Look at the pictures. What can you see? Talk about things that are blue. Colour the pictures blue.

Resources

flashcards, counters with blue objects and blue cubes, song: Colours
Teacher's Guide page 11

Four 4

Activity instructions

Talk about squares and measure squares to see that all four sides are equal. Colour the picture on the page to reveal the numeral 4. Colour the 4 butterflies.

Resources

flashcards, Colourful Shapes game, shape counters, tangrams, song: Number formations, Shape song
Teacher's Guide page 9

Four 4

1 2 3 4

4

Activity instructions

Top: Count the butterflies.
Middle: Trace the numeral 4.
Bottom: Find and colour the four butterflies. Colour the picture.

Resources

flashcards, frieze
song: Number formations

Four 4

4 four

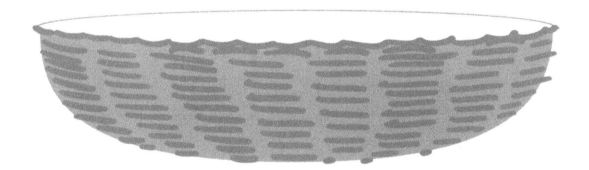

Trace	4	4	4	4		

Activity instructions
Top: Look at the numeral 4 and the number name.
Middle: Draw and colour 4 fruits in the basket.
Bottom: Trace the numeral 4.

Resources
flashcards, frieze, counters, cubes
song: Number formations

39

Four 4

HUGH

Trace	4	4	4	4		

Activity instructions

Top: Draw 4 lines on the roof.
Middle: Draw 4 flowers. Colour the house.
Bottom: Trace the numeral 4.

Resources

parrot puppet, flashcards, frieze
song: Number formations

Four 4

Activity instructions

Top: Count and colour the set of four. Trace the numeral 4.
Middle: Count and colour the set of four. Colour the numeral 4.
Bottom: Draw a set of 4 objects.

Resources

flashcards, frieze, counters, dice, cubes

4

Four 4

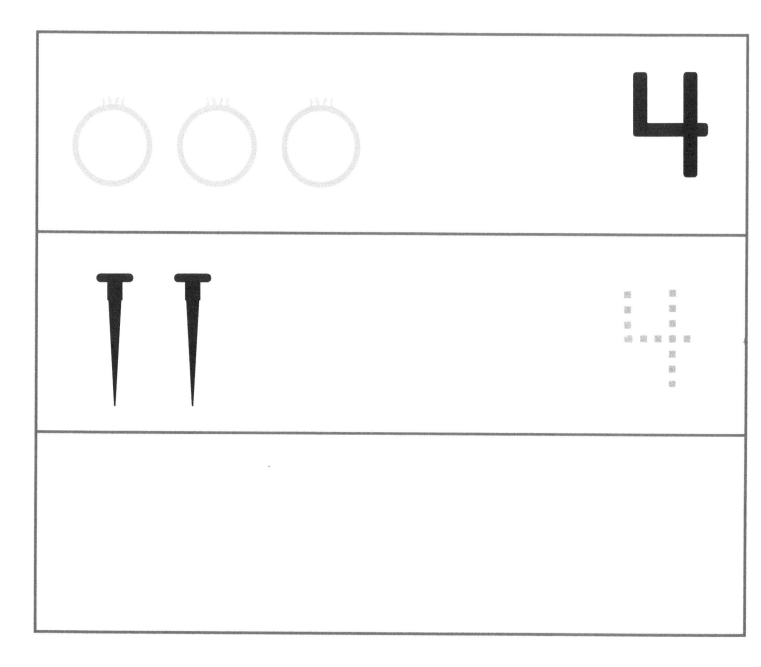

Activity instructions

Top: Draw one more ring.
Middle: Draw 2 more nails and trace the numeral 4.
Bottom: Draw 4 cookies. Trace the numeral 4 and the number name.

Resources

flashcards, frieze, counters, dice

Different

Activity instructions

Colour the umbrellas that are different from the umbrella on the box.

Resources

parrot puppet, counters, cubes
Teacher's Guide page 16

Hot and cold

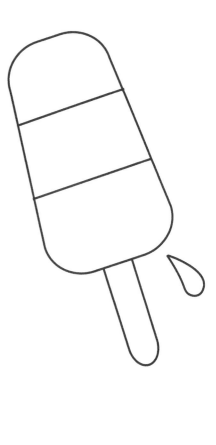

Activity instructions

Talk about how hot things are sometimes associated with red and cold things are sometimes associated with blue. Colour the pictures that are hot red. Colour the pictures that are cold blue.

Resources

flashcards, red and blue cubes, song: Colours
Teacher's Guide page 17

Five 5

Activity instructions

Colour the numeral 5. Colour or put glitter on the stars.

Resources

flashcards, frieze, dice, songs: Number formations;
Five little monkeys; Five little ducks
Teacher's Guide page 9

45

Five 5

1 2 3 4 5

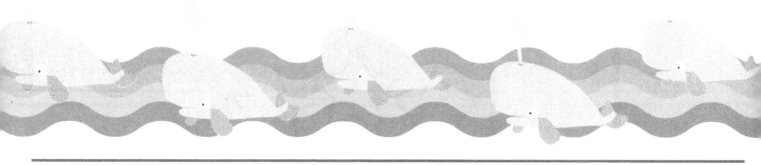

5 5 5 5 5 5 5

Activity instructions

Top: Count the whales.
Middle: Trace the numeral 5.
Bottom: Find and circle 5 whales. Colour the picture.

Resources

counters, cubes, dice, song: Number formations

Five 5

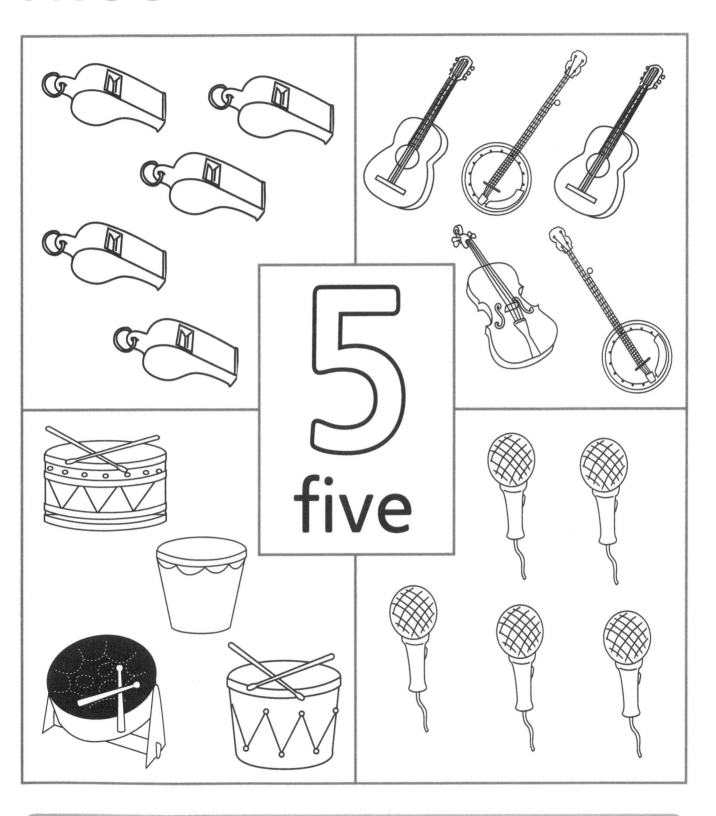

5

five

Five 5

On the fifth day of Christmas my true love gave to me, 5 golden rings.

Sara

Trace	5	5	5	5		

Activity instructions
Top: Draw 5 rings in Sara's trinket box.
Bottom: Trace the numeral 5.

Resources
parrot puppet, flashcards, frieze, counters, dice
songs: Number formations; Sally the camel has five humps.

Five 5

Here are 5 fingers plain to see. They are here to help you and me.

Activity instructions

Paint your palm and stamp it showing five fingers. Trace around your hand.

Resources

parrot puppet, flashcards, frieze, counters, dice
songs: Hokey cokey; Number formations

49

Five 5

Activity instructions

Top: Draw 1 more pak choi.
Middle: Draw 2 more beets.
Bottom: Draw 5 tomatoes. Trace the numeral 5 and number name.

Resources

flashcards, frieze, counters, cubes

More

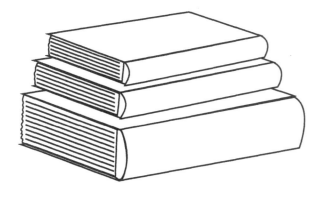 

Less

Activity instructions

Circle the sets with less objects. Colour the pictures.

Resources

counters, cubes
Teacher's Guide page 14

Hard and soft

Activity instructions

Talk about things that are hard and soft. Colour the pictures that are hard green. Colour the pictures that are soft yellow.

Resources

flashcards, frieze
Teacher's Guide page 17

Six 6

Activity instructions
Colour the numeral 6. Use the code to help you.

Resources
flashcards, frieze, counters, cubes, dice
songs: One, two, buckle my shoe; Number formations
Teacher's Guide page 9

54

Six 6

1 2 3 4 5 6

6 ́6 ́6 ́6 ́6 ́6

Activity instructions
Top: Count the frogs.
Middle: Trace the numeral 6.
Bottom: Circle 6 flies for Froggie's dinner. Colour the picture.

Resources
flashcards, frieze, dice, songs: One, two, buckle my shoe;
Number formations

55

Six 6

Activity instructions

Count and colour the sets of 6.

Resources

flashcards, frieze, counters, cubes, dice, Shape song
Teacher's Guide page 10

Six 6

Trace	6	6	6	6		

Activity instructions
Top: Draw 6 watermelons on the vine.
Bottom: Trace the numeral 6.

Resources
flashcards, frieze, counters, song: Number formations

Six 6

Uncle King's Farm

Trace	6	6	6	6		

Activity instructions

Top: Trace and colour 6 animals on Uncle King's farm.
Bottom: Trace the numeral 6. (Optional: cut out the animals from page 103 and stick them onto the farm.)

Resources

flashcards, frieze

Six 6

I found ☐ pigs with the numeral 6.

Six 6

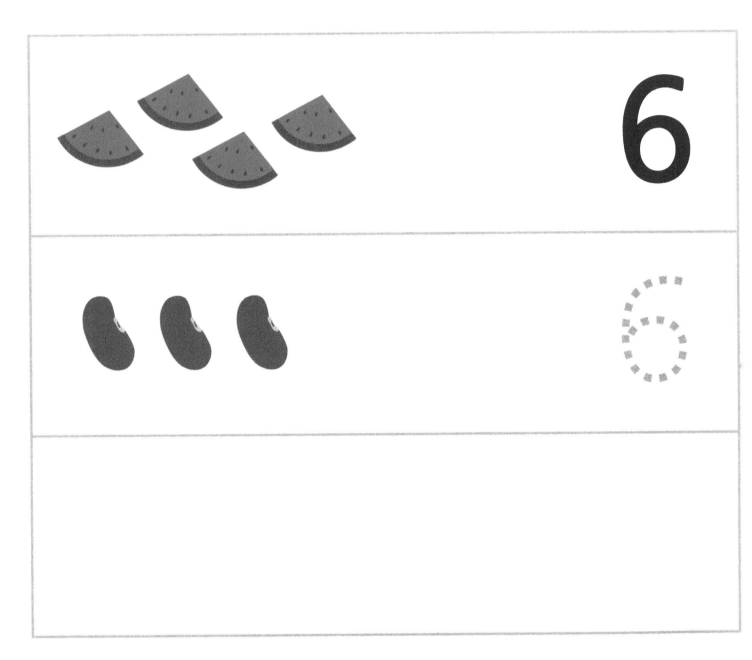

Activity instructions

Top: Draw 2 more pieces of watermelon.
Middle: Draw 3 more kidney beans to make 6.
Bottom: Draw 6 oranges. Trace the numeral 6 and the number name.

Resources

flashcards, frieze, dice, songs: One, two, buckle my show; Number formations

60

Looking back 👀

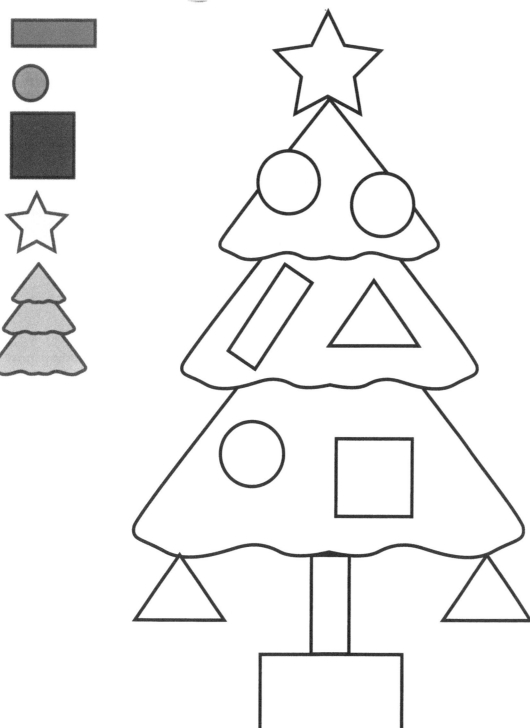

Activity instructions
Use the code to help you to colour the picture.

Resources
Colourful Shapes game, shape counters, Shape song
Teacher's Guide page 10

61

Looking back 👁 👁

5	
4	
3	

Activity instructions

Count the passion fruit in each bunch. Write the answer in the box.
Right: Look at the numeral. Draw circles to match the numeral.

Resources

flashcards, frieze, counters

62

Looking back 👁 👁

4 four 5 five	6 six 7 seven
2 two 3 three	4 four 5 five

Activity instructions
Top: Write the missing numerals on the dial.
Bottom: Count and circle the correct answer.

Resources
flashcards, frieze, dice, songs: One, two, buckle my shoe;
Number formations

Looking back

two 2

five 5

one 1

six 6

three 3

four 4

Seven 7

Activity instructions

Count and colour the bows on the kite's tail. Colour the numeral 7.

Resources

flashcards, frieze
Teacher's Guide page 9

65

Seven 7

1 2 3 4 5 6 7

7   7   7   7   7   7

Activity instructions

Top: Count the fish.
Middle: Trace the numeral 7.
Bottom: Sam is fishing. Count and colour 7 fish.

Resources
flashcards, frieze

Seven 7

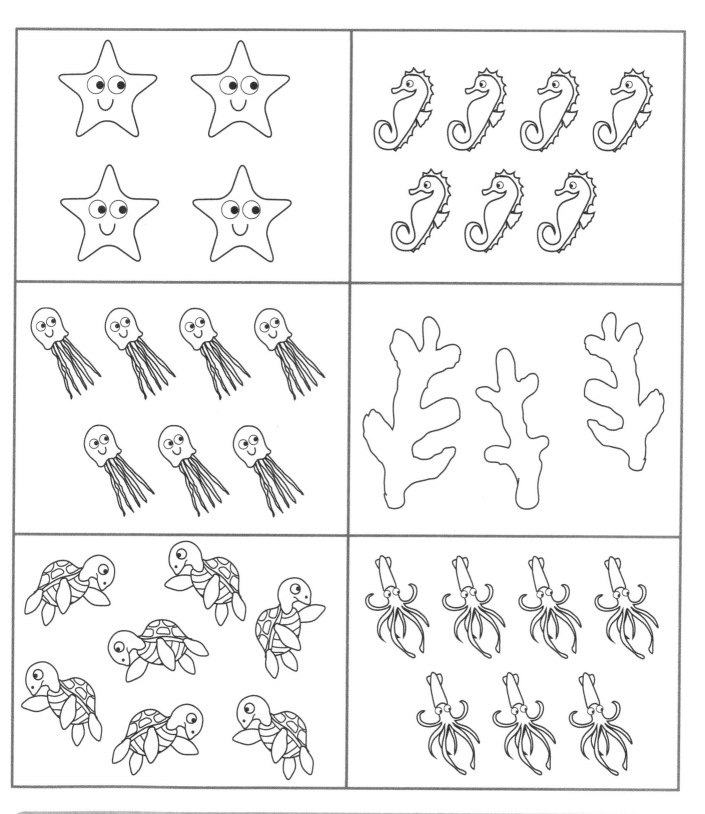

Activity instructions
Count and colour the sets of 7.

Resources
flashcards, frieze, song: One, two, buckle my shoe

Seven 7

Trace	7	7	7	7		

Seven 7

Trace	7	7	7	7		

Activity instructions
Top: Colour 7 spiders.
Bottom: Trace the numeral 7.

Resources
flashcards, frieze

Seven 7

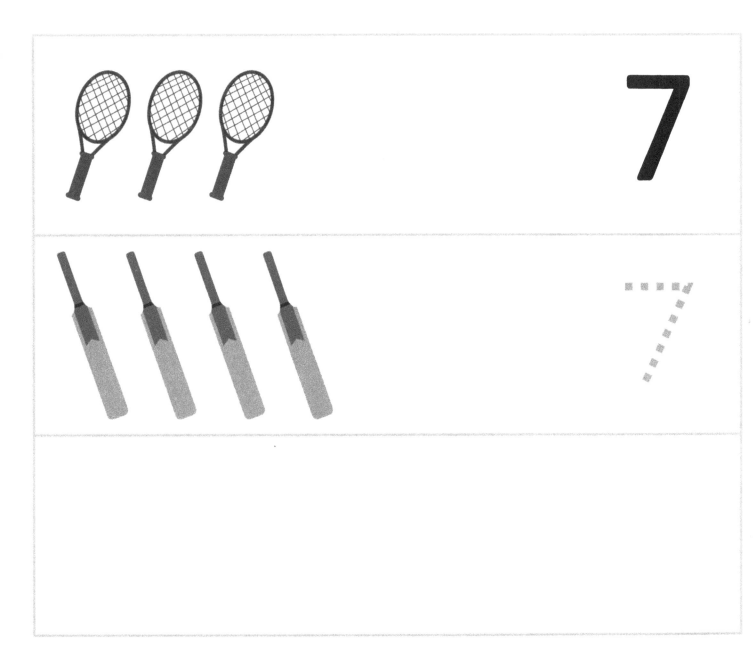

7

7

7 seven

Activity instructions

Top: Draw 4 more tennis racquets.
Middle: Draw 3 more cricket bats.
Bottom: Draw 7 footballs. Trace the numeral 7 and the number name.

Resources

flashcards, frieze, song: Number formations

Left and right

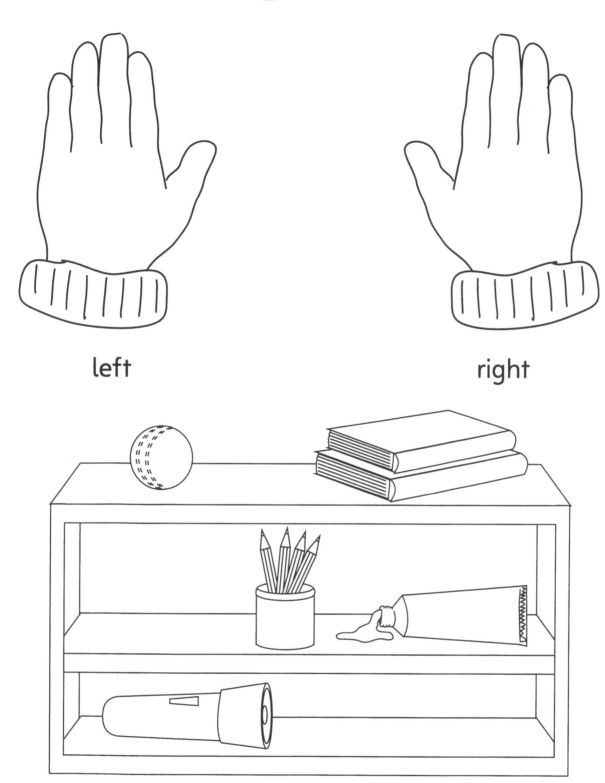

left   right

Activity instructions

Top: Colour the left and right hand.
Bottom: Circle objects on the left in red. Colour the objects on the right blue.

Resources

song: Hokey cokey
Teacher's Guide page 25

7

Picture graph

Favourite pets		

Activity instructions

Look at and talk about the picture graph. Colour the pet that is the favourite.

Resources

parrot, flashcards, frieze
Teacher's Guide page 20

Sorting

Activity instructions
Colour the △ , ▪ , ● , ▭ .

Resources
tangrams, Colourful Shapes game, shape counters,
Shape song
Teacher's Guide page 7

73

Empty

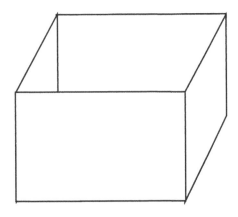

 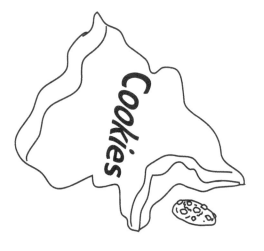

Activity instructions

Look at each set of objects. Circle and colour the ones that are empty.

Resources

story: Jack and Jill
Teacher's Guide page 19

74

Eight 8

Eight 8

1 2 3 4 5 6 7 8

8   8   8   8   8   8

Eight 8

Randy rabbit will eat ☐ carrots.

Activity instructions

Top: Help Randy Rabbit to eat his carrots! Colour the carrots so he can see them.
Bottom: Write the missing numeral in the box.

Resources

flashcards, frieze, counters

Eight 8

8

eight

Trace	8	8	8	8		

Activity instructions

Top: Colour the nest with 8 eggs.
Bottom: Trace the numeral 8.

Resources
flashcards, frieze

Eight 8

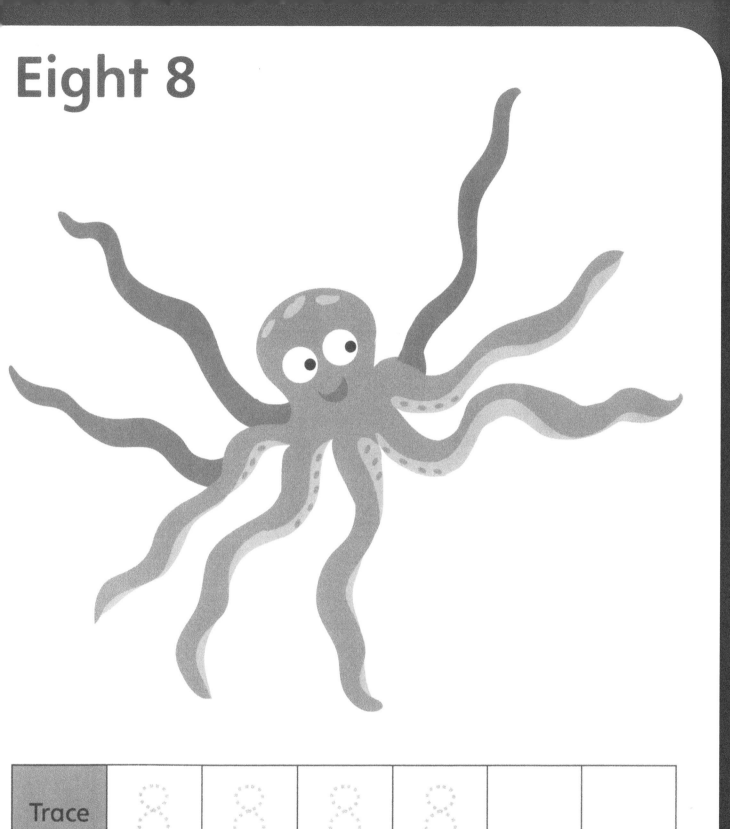

Trace	8	8	8	8		

Activity instructions

Top: Ollie Octopus likes lollipops. Draw 8 lollipops for Ollie.
Bottom: Trace the numeral 8.

Resources

flashcards, frieze, song: Number formations

Eight 8

8

Top: Draw 5 more leaves.
Middle: Draw 4 more leaves. Trace the numeral.
Bottom: Draw more leaves to make 8. Trace the numeral 8 and the number name.

flashcards, frieze, song: Number formations

Position

Activity instructions

Talk about the picture and the things that are above, on, in, and under the birdbath. Colour the bird that is above the butterfly red. Colour the bird that is in the water green.

Resources

parrot puppet
Teacher's Guide page 21

Sorting

Nylah's Toy Box

Activity instructions

Help Nylah to tidy her toys.
Draw a line from the toys to the toy box.
Colour all of the toys.

Resources

parrot puppet
Teacher's Guide page 7

Matching

Activity instructions	Resources
Match each paint brush to its can of paint by drawing a line.	different coloured cubes Teacher's Guide page 7

Nine 9

Activity instructions

Look at the planets. How many can you see? Colour the numeral 9.

Resources

flashcards, frieze, songs: Number formations; One, two, buckle my shoe
Teacher's Guide page 9

Nine 9

1 2 3 4 5 6 7 8 9

9 9 9 9 9 9

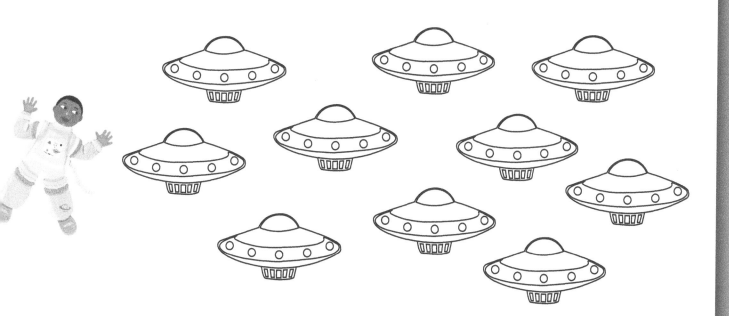

Activity instructions
Top: Count the rockets.
Middle: Trace the numeral 9.
Bottom: Joe is going to the moon. Colour 9 spaceships.

Resources
flashcards, frieze, song: Number formations

Nine 9

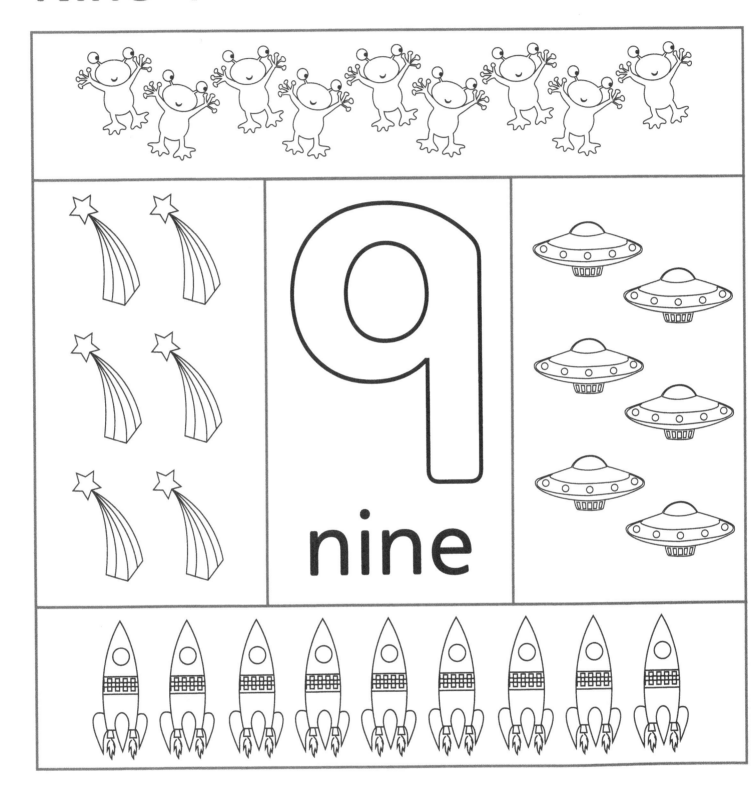

nine

Activity instructions

Look at the objects. Count and colour the sets of 9.

Resources

flashcards, frieze

Nine 9

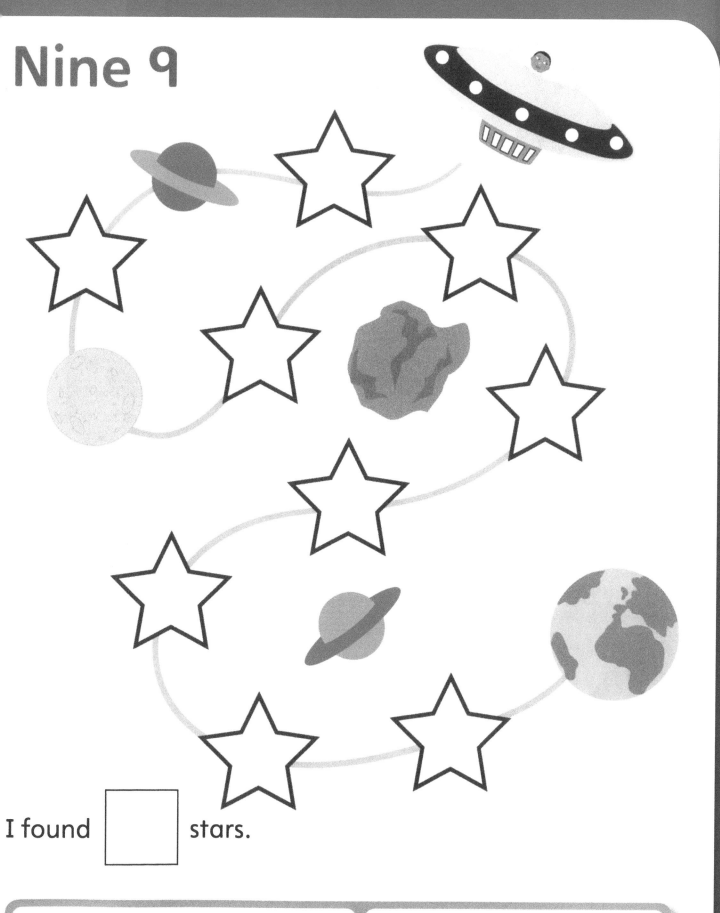

I found ☐ stars.

Activity instructions

Look at the picture. Count the stars. Write the missing numeral in the box. Help Joe to find his way home by colouring all the stars.

Resources

flashcards, frieze

Nine 9

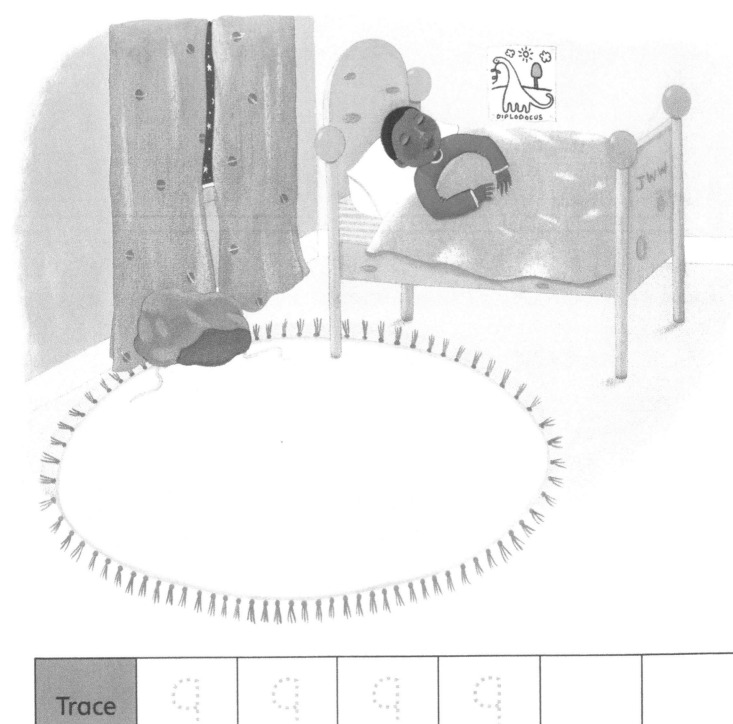

Trace	9	9	9	9		

Activity instructions

Top: Joe took home some moon rocks. Draw 9 rocks on the mat.
Bottom: Trace the numeral 9.

Resources

flashcards, frieze, songs: Number formations; One, two, buckle my shoe

Nine 9

9

Activity instructions
Top: Draw 5 more suns.
Middle: Draw 4 more aliens. Trace the numeral.
Bottom: Draw 3 more clouds. Write the missing numeral.
Trace the numeral 9 and the number name.

Resources
flashcards, frieze, song: Number formations

Colour the numeral

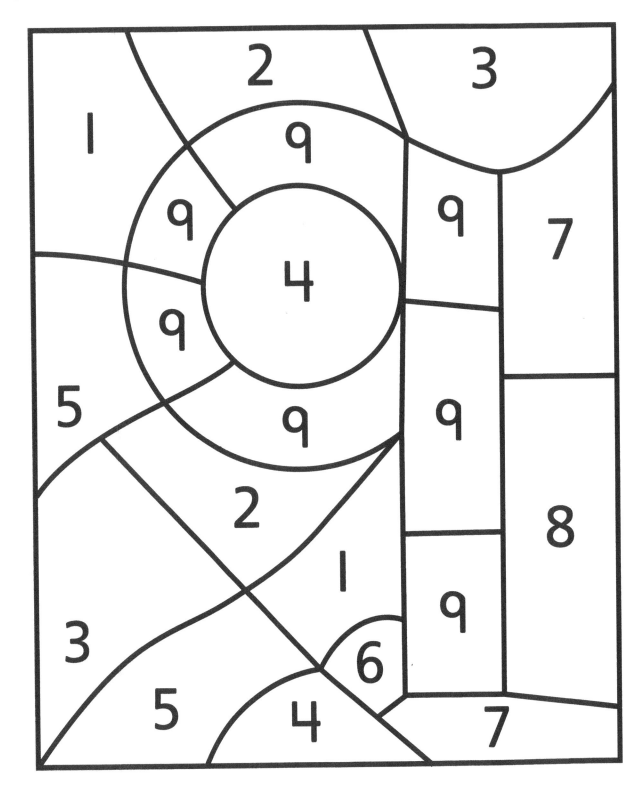

Colour the spaces showing the numeral 9 with the same colour.
Colour the rest of the picture. Use a different colour for each numeral.

Resources
flashcards, frieze, cubes
Teacher's Guide page 11

Ten 10

Trace

Activity instructions

Top: Colour the numeral 10. Count and colour the helicopters.
Bottom: Trace the numeral 10.

Resources

flashcards, frieze, songs: Number formations; One, two, buckle my shoe; Ten green bottles
Teacher's Guide page 9

Ten 10

1 2 3 4 5 6 7 8 9 10

10   10   10   10   10

Activity instructions

Top: Count the balloons.
Middle: Trace the numeral 9.
Bottom: Find and circle 10 airplanes.

Resources

flashcards, frieze, songs: Number formations; Ten green bottles

What's different?

Activity instructions	Resources
Look at the two pictures. Find and circle 3 things that are different in the bottom picture.	flashcards, frieze Teacher's Guide page 16

93

Time

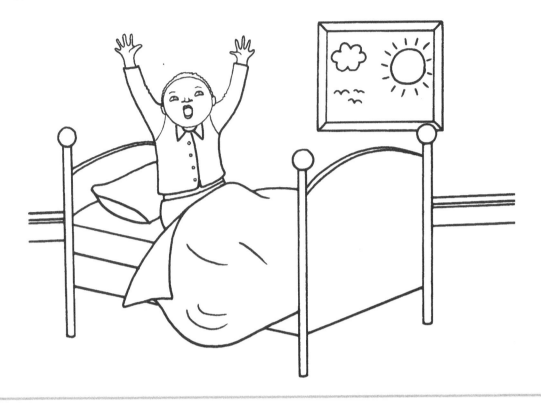

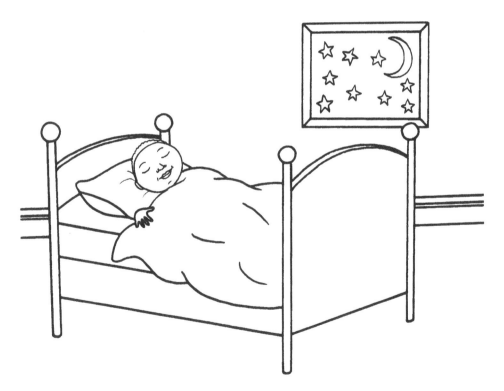

Activity instructions

Talk about the two pictures. Colour the picture that shows day.

Resources

flashcards, frieze
Teacher's Guide page 27

Matching

Activity instructions

Match a hat and vest to each teddy bear. Use the shapes to help you.

Resources

tangrams, shape counters, Colourful Shapes game, Shape song
Teacher's Guide page 7

Ten 10

Trace	10	10	10	10		

Ten 10

Trace	10	10	10	10		

Activity instructions

Top: Khyle is picking mangoes. Can you help Khyle? Draw 10 mangoes beside his bucket.
Bottom: Trace the numeral 10.

Resources

flashcards, frieze, songs: Number formations; Ten green bottles

Ten 10

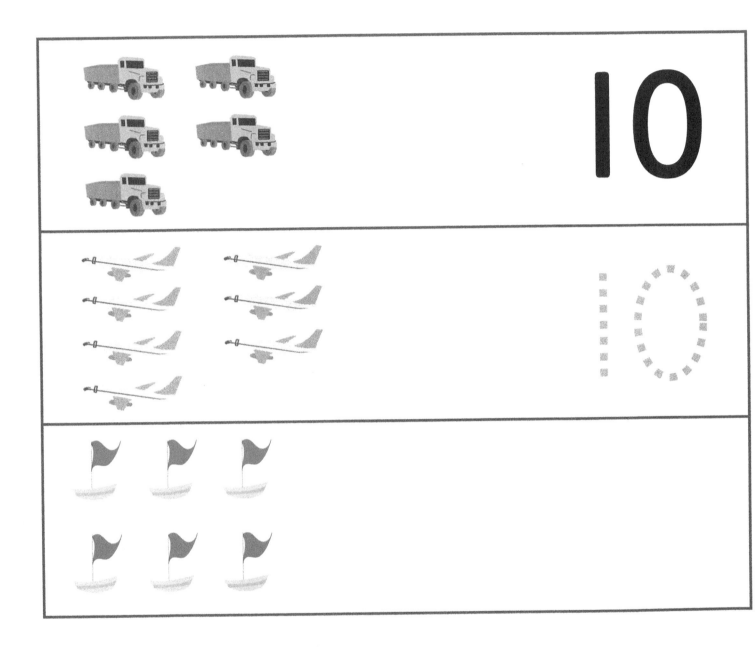

10

Activity instructions

Top: Draw 5 more trucks.
Middle: Draw 3 more airplanes. Trace the numeral.
Bottom: Draw 4 more boats.
Trace the numeral 10 and the number name.

Resources
flashcards, frieze, song: Number formations

Ten 10

Trace	10	10	10	10		

Activity instructions
Top: Smiley has lost some of his balls. Draw more balls to make 10.
Bottom: Trace the numeral 10.

Resources
flashcards, frieze

99

Looking back

Activity instructions

Help Nick to find his dog. Count the flowers along the way. Write the answers in the boxes.

Resources

flashcards, frieze
Teacher's Guide page 8

Looking back

Activity instructions

Top: Write the missing numerals on the dial.
Bottom: Draw objects in each circle to match the numeral.

Resources

flashcards, frieze, counters, cubes

101

Looking back

 10

 7

9

 8

Activity instructions
Count the objects in each row. Draw a line to the correct numeral.

Resources
flashcards, frieze, counters, cubes

Six 6 (optional activity)

Activity instructions

Colour and cut out the animals. Stick them onto Uncle King's farm on page 58.